The Basic Essentials of
WEATHER
FORECASTING

by Michael Hodgson

Illustrations by
Devin Wick

ICS BOOKS, Inc.
Merrillville, Indiana

THE BASIC ESSENTIALS OF WEATHER FORECASTING

Copyright © 1992 Michael Hodgson

10 9 8 7 6 5 4 3 2 1

Printed in U.S.A.

DEDICATION

This book is for my wife, Karen, who has patiently shown me through the years that, no matter what the weather condition, it's always warmer in the tent.

Published by:
ICS Books, Inc.
One Tower Plaza
107 E. 89th Avenue
Merrillville, IN 46410
800-541-7323

Library of Congress Cataloging-in-Publication Data

Hodgson, Michael.
    The basic essentials of weather forecasting / by Michael Hodgson ; illustrations by Devin Wick.
        p. cm. -- (The Basic essentials series)
        Includes index.
        ISBN 0-934802-75-0 : $4.95
        1. Weather forecasting--Popular works.   I. Title.   II. Title: Weather forecasting.   III. Series.
    QC995.43.H63   1992
    551.6 ' 3--dc20   92-7829
                                                           CIP

TABLE OF CONTENTS

INTRODUCTION v

1. HOW WEATHER HAPPENS 1

2. UNDERSTANDING CLOUDS
   AND WHAT THEY MEAN TO
   THE WEATHER OBSERVER 13

3. GEOGRAPHIC WEATHER
   VARIATIONS 23

4. FORECASTING CHANGES
   IN WEATHER 34

5. NATURE'S SIGNS—
   Semi-Reliable Forecasting
   From Legends and Lore 40

6. BACKYARD METEOROLOGY 47

APPENDIX I 53
Glossary of Terms

APPENDIX II 57
Recommended Reading

APPENDIX III 59
Suppliers of Weather Instruments

APPENDIX IV **61**
Weather Rules of Thumb

INDEX. **63**

INTRODUCTION

"Tonight's weather is dark, followed by widely scattered light in the morning..."
George Carlin from his Hippi Dippi Weatherman routine.

A number of years ago, actually more years than I care to remember, two friends and I were backpacking along the Appalachian Trail, just north of Great Smoky Mountain National Park. We were on a week-long backpack to celebrate the independence of our 16-year old spirits.

Just after dawn on the second day of our adventure, the clouds which had been stacking ominously for hours released a deluge of water lasting for several hours. The deluge reduced to a steady drizzle that continued until evening. Soaked and somewhat dispirited, we trudged into camp, happening upon a crusty old traveler slouched against a pack that appeared as if it may have been a prototype for later cruiser-frame models.

I glanced pensively up at the lowering ceiling, which would soon turn to a dense mist, and then over to the old-timer who had either been ignoring our dripping entrance or was unaware of our presence—I sensed the former. Shuffling closer, not wanting to offend, I managed a somewhat unsure, "Mister, is this weather going to keep up or do you think it'll be sunny tomorrow."

With a cough, the old-timer lifted the brim of his felt hat, gazed intently at me and my friends for what seemed an eternity and then, without smiling, drawled, "Waell boys. I cin garantee ya'll one thing fer sher. Come mornin' there's bound to be weather of some sort or the other. Wet or dry, ya'll don't got much choice in the matter so why waste yer time frettin over what ya cain't control." With those few words of wisdom, the old-timer pulled himself upright, swung his pack onto one shoulder, tipped his hat respectfully in our direction and then disappeared into the dusk and mist.

I'll never forget those words and though, for the most part, they are filled with truth, there is one important element missing. While one cannot do much about the weather, by learning to read and understand changes in weather patterns and what those changes mean, one can experience the vast difference between blind reaction and reliable preparedness. Often, that difference alone may determine the margin of comfort and safety that separates disaster from adventure.

Continually practice keeping a weathered eye turned upwards towards the sky. The more you are aware and the more you learn about what causes weather, the more perceptive will be your observations and the more accurate your guesses as to the weather's outcome. But, never forget — predictions relative to weather are only educated guesses, never statements of fact. Always be prepared for the worst.

1. HOW WEATHER HAPPENS

In the Northern Hemisphere, warm air (tropical) moves north and cold (polar) air south, generally speaking. With that in mind, one can expect warm fronts to be generated from the southern reaches and cold fronts from the northern. On weather charts and some maps, three types of air mass are often noted—Maritime Polar (mP), cold polar air that formed over the ocean; Maritime Tropical (mT), warm tropical air that

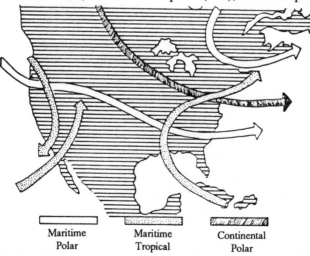

| Maritime | Maritime | Continental |
| Polar | Tropical | Polar |

Figure 1-1 Movement of air masses, Maritime Polar, Maritime Tropical, Continental Polar.

1

formed over the ocean; Continental Polar (cP), cold polar air that formed over land.

With polar air masses, the weather is apt to change abruptly and as the air warms over land, it becomes turbulent, with associated cumulus clouds and often heavy precipitation. Tropical air is more stable, since it is already quite warm, and while it often brings precipitation, the weather associated with a tropical air mass is apt to stay around for a while.

Both continental and polar air influence local weather conditions around North America. In San Francisco, California, it is the cold maritime polar air during the summer that causes coastal fog and in the winter heavy rains. Sometimes, in the summer, thundershowers form in the Sierra dropping precipitation on the western slopes.

Continental polar air causes turbulent weather conditions and rains to fall in the Great Lakes region during the summer and heavy snows in the southeastern reaches of the lakes in the winter.

Maritime tropical air brings with it humidity and extreme heat to the East during summer months. In the winter it brings very heavy rains.

As an air mass moves around the earth's surface, contact with other air masses is inevitable. These points of contact are called fronts and understanding how fronts interact is critically important to understanding how weather happens.

Fronts don't just suddenly appear out of nowhere, however. Forces, called pressure systems, are at work, pushing and pulling the various cold and warm air masses at will. High pressure cells of the Northern Hemisphere create winds that rotate in a clockwise direction while winds associated with low pressure cells rotate counterclockwise. The reverse is true south of the equator

High pressure systems are associated with relatively cold air while low pressure systems are associated with warm air. Since cold air is heavier than warmer air, it exerts a downward pressure on the earth. This pressure is registered by a rise in barometric pressure on a barometer. Conversely, warm air rises, resulting in a releasing of pressure on the earth's surface. This pressure is registered by a lowering in barometric pressure on a barometer.

Cold and warm fronts never mix. They displace each other, forming associated,but separate systems that are independent of one another and, in actuality, alternate with each other. The weather observer can determine changes in weather patterns by noting changes in barometric pressure, coupled with prevailing winds. More on how to do that in Chapter Four.

Although wet and turbulent weather tends to be associated with low pressure systems, or depressions, and fair weather with high pressure systems, or ridges, exceptions to the rule do happen. The weather associated with either a cold or warm front is just as likely to be fair, or poor, when either is in firm control, depending on the direction of the winds brought by the prevailing system. If the winds have passed over a significant body of water, either system, high or low, can bring precipitation.

How quickly barometric pressure changes as a particular system moves in is also an indication as to how long in duration and/or how severe any accompanying storm may be. A rapid drop in barometric pressure indicates that any storm brought in by the low pressure system will likely be short. If the drop in pressure is slow and steady, however, expect the accompanying storm to be long and severe.

Rising pressure usually brings a fair change to the weather. If, however, the pressure was very low to begin with, it is possible for a period of intense rain squalls to move through before any sunshine dances among the leaves. A rapid rise in pressure will often bring high winds due to unstable atmospheric conditions.

Fronts

When air masses of different temperatures and densities meet each other, they do not mix well. Think of a bunch of punk rockers crashing a Frank Sinatra benefit—there's bound to be tension. When this happens, battle lines are drawn and a front is defined. There are three types of fronts: warm, cold, and occluded.

Warm Fronts

Warm air is far more stable than cold air. It is also more moist with lower ceilings and poorer visibility, even if there is no appreciable precipitation. While weather associated with warm fronts is typically less severe that those attributed to cold fronts, the weather is frequently longer lived — rain may last several days or more.

Warm fronts move at a relatively slow speed, 10 to 20 miles per hour, and may be anticipated as much as two days in advance by a consistent sequence of cloud formations and a drop in barometric

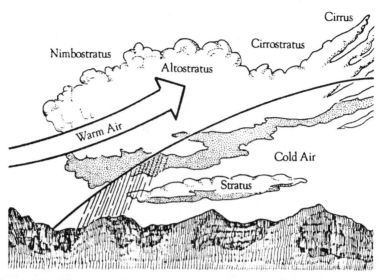

Figure 1-2   Warm Front Cloud progression and air mass flowing over the ground.

pressure (altimeter needle gains elevation). The usual cloud formation announcing an incoming warm front is preceded by cirrus clouds followed by, in succession, cirrostratus, altostratus, and finally nimbostratus.

As a warm front moves in, it slowly displaces the cold air that came before by rising above it and slowly warming it. The gentle slope of the warm front, as it is forced up and over the cold air, gradually cooling to dew point, results in the predictable formation of clouds as described in the preceding paragraph.

Cold Fronts

Cold air is more unstable than warm air and consequently very active. High ceilings and good visibility are associated with cold fronts, unless there is precipitation. Weather associated with a cold front is often severe, violent in nature. Typically, though, weather conditions associated with cold fronts are also shorter in nature than those associated with warm fronts. Cold fronts move at a speed of approximately 25 to 35 miles per hour and generally originate in the north or west when forming in the Northern Hemisphere.

As a cold front moves in, it pushes under the warm air, which rises and cools. If dew point is reached by this rising and cooling, precipitation occurs, often quite heavily. Cold fronts frequently arrive with very little warning. Nimbostratus and cumulostratus are both rain clouds that may be generated by a cold front. Both are usually preceded by altostratus, possibly after an advance squall line of thundershowers has passed through.

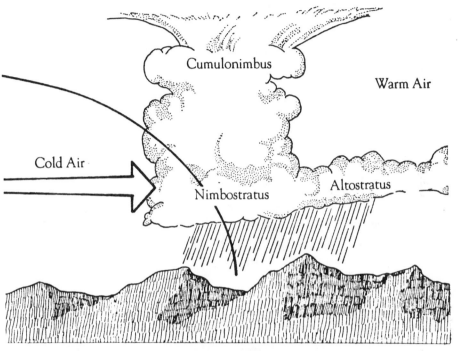

Figure 1-3   Cold Front Cloud progression and air mass flowing over the ground.

Occluded Front

An occluded front occurs when one air mass gets caught between two other air masses and forced off the ground. What actually happens depends almost entirely on temperature. As a warm front moves through, with a cold air mass in front and a cold front behind chasing it, several things can happen. If the air behind the pursuing cold front is colder than the air in front of the warm air mass, then the advancing cold front will actually lift both the warm and cold air in front of it. Weather associated with this type of front is usually squalls with thunder and lightning.

If the temperatures are reversed, with the colder air in front of the warm air, the pursuing cold front will be forced up over the warm front. The weather is likely to include heavy precipitation.

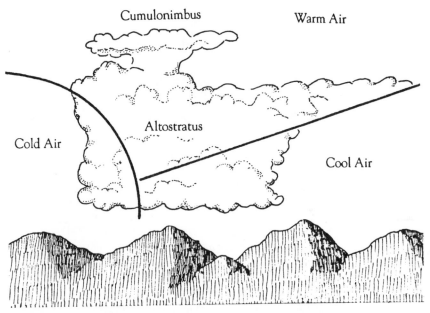

Figure 1-4 Occluded Front Cold front lifting both cold and warm front off the ground.

Thunderstorms and Lightning

Extreme and violent vertical movements of air, updrafts, often lead to thunderstorms. This uplifting usually creates dramatic cloud formations, cumulus and cumulonimbus, that rise vertically, sometimes as high as 70 to 75 thousand feet. The uplift of air can be caused by heating of the air from the ground (common in the Midwest), weather associated with a cold front, or temperature differences caused by the meeting of land and sea (common in the Gulf and southern Atlantic states).

Because temperature differences are the most common cause of thunderstorms, thunderstorms usually occur in the afternoon, when temperature differences between land and air or sea are the most extreme. Since temperature differences between land and sea are not as dramatic in the north Atlantic or West Coast (Pacific Ocean) regions, thunderstorms in these locations are not as frequent.

A thunderstorm forms as rising warm air cools and condenses. The higher the cloud stacks, the greater the level of cooling and soon precipitation begins. As the falling rain and ice crystals cool the air within the cloud, the temperature difference between the cloud and the surrounding air begins to equal, leading to the creation of downdrafts and heavy precipitation. The rush of cold air fans out ahead of the storm, sometimes as much as two to three miles, and is a good indicator of an impending thunderstorm

As the downdrafts increase in intensity, the cloud becomes nothing more than downward moving air. Cooling of air ceases and since no air is rising and condensing, the rain tapers off.

Lightning is caused by the attraction of unlike electrical charges within the storm cloud or the earth's surface. The friction caused by rapidly moving air particles, churning from the violent up and downdrafts of air, leads to a building up of strong electrical charges. As the electrical pressure builds, charges between parts of the cloud or from the cloud to the earth are released, taking the form of lightning.

Up to 30 million volts can be discharged by one single lightning bolt and it is this power, or explosive heating of the air that causes the compressions of thunder.

It is possible to roughly judge the distance of an approaching storm by observing the lightning's flash followed by the resounding boom of thunder. For this method, count slowly, "one-one thousand, two-one thousand, three-one thousand" and so forth. This will approximate one second elapsed for each thousand counted. Once the lightning has flashed, begin to count. When you hear the thunder, stop counting. Every five seconds of elapsed time indicates one mile of distance. In other words, if the count reaches "seven-one thousand", it is safe to assume that the approaching thunderstorm is approximately one and a half miles distant.

Thunderstorms are a dangerous companion to have when traveling exposed on a mountain peak. If a thunderstorm approaches when you and your hiking party are on an exposed peak or ridge, take the following precautions:

- Get off the ridge or peak if at all possible. Even a few feet below the highest point around you is better than not moving at all.
- Get away from your pack, as the metal in the pack will conduct electricity.
- Position yourself on a dry surface, preferably insulated such as a

Figure 1-5 Cone of Protection

sleeping pad. If the sleeping pad has become soaked in the rain, however, it will be useless as water conducts electricity.

- Do not lie down or sit. Rather crouch, with your feet close together. The idea is to minimize the available surface area through which possible ground currents from nearby lightning strikes may move.

- Do not huddle next to a single tree. That is rather like hugging a lightning rod and expecting to be safe. Choose a cluster of trees instead and place yourself in the middle, preferably in an open area.

- Spread the group out, at least 25 to 30 feet apart. If lightning does strike, the idea is to minimize the potential damage and injury. With the group spread out, chances are only one person will be injured, if at all, and this leaves the rest of the party to provide life-saving assistance once the storm moves on.

- Avoid any depressions or caves. Such areas usually have a presence of moisture in them which makes them more susceptible to conducting electricity.
- If you are caught on a lake in a thunderstorm, assuming a crouching position towards the middle of the boat. Try to minimize your contact with wet objects.

Hurricanes

Hurricanes are tropical cyclones, low pressure systems. Fortunately, the typical outdoorsman, or woman, has little to worry about from hurricanes due to their rarity of occurrences in recreational areas away from the ocean. Hurricanes only form over open ocean areas that are covered by an extremely warm and moist air mass— which is why they are usually called tropical cyclones. Hurricanes will always break up and lose intensity within hours of moving over a large land mass. Once inland, the major effect that is felt from a hurricane is near torrential downpours, but not, thankfully, the severe winds felt in coastal regions. It is virtually impossible for the amateur weather observer to anticipate the approach of a hurricane, other than listening to the National Weather Service. The associated clouds and lowering of atmospheric pressure that precede the arrival of a hurricane is decidedly similar to that of a warm front.

Tornados

Tornados are perhaps the most violent and intense of all known storms, with winds in excess of 300 miles per hour recorded in the vortex. Man-made structures seem to literally explode as the extremely low pressure within the vortex of a tornado causes the normal pressure trapped within structures to expand rapidly, ripping buildings to shreds from the inside out.

Violent updrafts, recorded between 100 to 200 miles per hour, within the center of the funnel cloud have been known to suck anything within its path hundreds of feet in the air before hurling them some distance away. Over bodies of water, funnel clouds lift water into the air creating what is known as a water spout.

Rain and Snow

Not every cloud produces precipitation, and even in those clouds that do, the conditions must be just right for rain or snow to fall. Every cloud is made up of millions of tiny droplets of moisture or ice crystals, so small in fact that each droplet or crystal is easily held aloft by moving air even though the pull of gravity on them is constant. The only way each droplet or crystal can fall to earth is if it gains enough size so that the pull of gravity overcomes the lift of moving air. This process of growth is referred to as coalescence.

Coalescence can occur in a variety of ways. The two main methods of formation are described as follows: 1. Droplets of rain collide, becoming bigger. As they become bigger, they are not whipped around quite so much by the movement of air and because of a larger surface area, collide with more and more droplets until they become large enough to fall as rain. 2. When water droplets and ice crystals occur in the same cloud, as they do in cumulonimbus, some of the water will evaporate and then condense again on the ice crystals. As the crystals grow, they fall towards earth as snow or ice pellets that melt in the warmer air of lower elevations, finally splashing down as raindrops.

Snow forms only when supersaturated air in the cloud is cold enough, at or below 10 degrees F, and when there are sufficient small particles of debris in the cloud upon which water vapor can crystallize. If supersaturated air within the cloud is colder than -38 degrees F, snow can form without the presence of debris.

Dew and Frost

Dew is not precipitation. It is, rather, water vapor that condenses on the surface of solid objects (such as a tent, grass, pack) that have cooled below the condensation point of the air that is in contact with them. Dew is usually heavier nearer bodies of water, in valley bottoms near streams or rivers, and on clear nights when cooling by radiation is at its greatest.

Frost is a type of dew, but it only forms when surface temperatures are at or below freezing. Frost forms as water vapor crystallizes instead of condensing into water droplets.

The Seasons

The seasons are created by the earth's orbit around the sun coupled with the fact that the earth's rotation on its axis is angled at 23.4 degrees to the plane of its orbit. Say what? It's not as complicated as it sounds. First, the earth is constantly spinning, one rotation every 24 hours, on an axis—think of a globe. Note how every globe is angled, 23.4 degrees. The earth also orbits the sun, one rotation every 365 days. The orbit is maintained on a constant plane, which means that the earth's

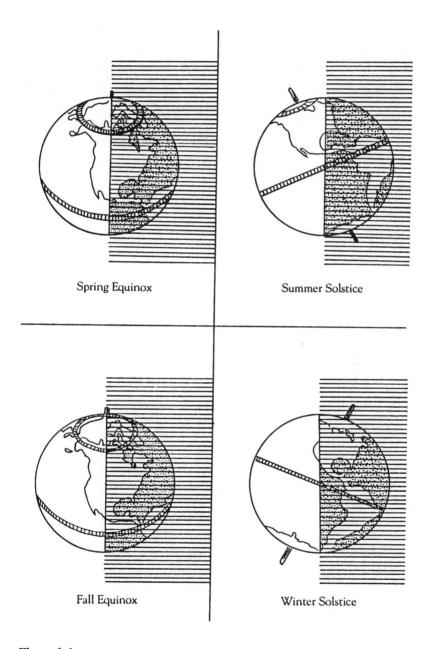

Figure 1-6

relative position to the sun take on three separate angles to the sun as it maintains its orbit.

In the Northern Hemisphere summer, the north pole is angled towards the sun. More sun is hitting the northern reaches of the globe meaning days are longer and the sun's rays are striking the earth more directly in the north. This is why summer is warmer than winter. In the winter, the south pole is angled towards the sun which reduces the sun's contact with the earth in the Northern Hemisphere and also diffuses the sun's warming rays, effectively lowering temperatures. In the spring and fall, the north and south poles are relatively equidistant from the sun. The seasons are reversed in the Southern Hemisphere.

2. UNDERSTANDING CLOUDS AND WHAT THEY MEAN TO THE WEATHER OBSERVER

Water Cycle

Although not specifically an integral part of weather forecasting, the water cycle does play a vital role in the formation of clouds and the eventual precipitation that returns the moisture to earth. Moisture in the air comes from many sources. Water evaporates into the atmosphere

Figure 2-1   Hydrological cycle

from oceans, seas, lakes, rivers, ponds, wetlands—any open body of water that comes into contact with air. As this warm, moisture-laden air cools, either adiabatically or by coming into contact with cooler air, the water vapor in the warm air condenses, forming clouds and then, inevitably, falls back to earth as precipitation—rain, snow, sleet, hail, etc.

It is also important to realize that all living things, plants and animals, give off water vapor. Animals give off water vapor through breathing, plants through their leaves. This moisture also contributes to the water cycle, which is an endless process of water being evaporated, condensed, evaporated, condensed, and on, and on.

Cloud Formation

In general, cloud formations are made up of water droplets and ice crystals that are held aloft by turbulent air movements. As the droplets of moisture or crystals of ice within clouds increase both in number and size, they become too heavy for the air to support and fall to the earth as precipitation. These water droplets and ice crystals all got where they are the same way, by the cooling of air below its saturation point, or the point at which water vapor condenses.

Clouds can be formed any number of ways, but the most common are as follows: 1. As the earth radiates heat during a clear night, fog may form near the earth's surface due to warm air contact with a cooling earth. 2. As a cold front pushes under a warm front, clouds form as the warm air, now cooling as it rises, condenses. 3. Coastal areas are very familiar with the fog and mist caused by warm ocean air moving over a cooler land surface and condensing. The same conditions can be associated with any large body of water. 4. Precipitation falling from high clouds sometimes cool the air below causing lower elevation clouds to form. 5. Clouds form as air cools and warms adiabatically from being pushed over a mountain ridge or peak.

What Do Clouds Mean?

Through the consistent observation of clouds, the weather watcher can draw some relatively accurate conclusions regarding shifting weather patterns. If clouds are massing and generally increasing in size and density, the weather is possibly changing for the worse. As clouds begin to move faster across the sky, this would indicate a change in wind velocity and also a change in pressure indicating an approaching storm.

Since varying cloud formations take on different appearances and altitudes, it is possible, although sometimes difficult, to observe the movements of two, or more, different cloud formations across the sky. Cloud formations that are moving in separate directions and at different altitudes often announce an impending storm.

Small, black clouds often bring rain. Scud clouds, small, dark cumulous clouds, that are seen sweeping beneath a dark or darkening stratus layer usually indicates an impending storm with accompanying winds.

Finally, observe cloud movements and wind direction together. One may assume that if both the cloud movement and wind direction are the same, that any weather will come from the direction the wind is blowing.

Types of Cloud Formations

Cirrus Clouds

Cirrus: Made up predominantly of ice crystals, cirrus clouds often are associated with other cloud formations, especially cirrocumulus.

Figure 2-2 Cirrus clouds

Cirrus clouds are arguably some of the most beautiful of the cloud formations, leaving milky white swirls and curls etched across the sky. Mares tails are a type of cirrus cloud.

Cumulus Clouds

Cumulus: Often referred to as heap clouds, cumulus clouds are, essentially, typified by heaped or fluffy formations.

Cirrocumulus: High level (usually around 25,000 ft.) heap clouds. Very often seen combined with cirrus clouds. Cirrocumulus indicates a condition of unstable air and may lead to precipitation before long.

Figure 2-3   Cirrocumulus

Altocumulus: Medium level (usually hovering around 8,000 ft.) fleecy or puffy clouds, similar to dense cirrostratus, but without any telltale halo. When viewed in the early morning, altocumulus usually indicates thunderstorms or precipitation within 24 hours—often that afternoon.

Figure 2-4   Altocumulus

Cumulonimbus: Often massive cumulus with a broad base ranging from 3,000 ft. upwards to 16,000 ft—even 65,000 ft. is not unusual. Top is fuzzy or anvil shaped. Heavy downpours, coupled with hail, lightening and thunder are standard fare.

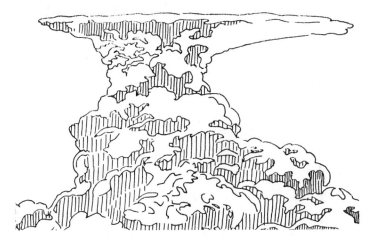

Figure 2-5   Cumulonimbus

Fair-weather cumulus: Low level (up to 4,000 ft) heap or cumulus clouds that often form in the late morning or early afternoon. Clouds are not very dense, are white in color, and well separated from each other. These clouds form when the air mass is stable and being warmed by the earth's surface.

Figure 2-6   Fair weather cumulus

Swelling cumulus: Medium level (around 15,000 to 20,000 feet) heap clouds that also form in the late morning or early afternoon. Weather observers often refer to these clouds as giant heads of cauliflower because their bottoms are flat and their tops are bumpy. Swelling cumulus indicates unstable air masses. These clouds are quite common in the deserts of the Southwest.

Figure 2-7   Swelling cumulus

Cumulus congestus: High level (stacked from 5,000 up to 30,000 feet) heap cloud formed by massive uplifting of heated air within a very unstable air mass. Its top is still bumpy and forming, marking the major difference between this cloud and the precipitating cumulonimbus.

Figure 2-8   Cumulus congestus

Stratus

Stratus: Stratus means layered, essentially formless with no real defining base or top. Fog is a type of stratus cloud which lies close to the ground and is caused when the Earth's surface cools. This cooling effectively lowers the air temperature resulting in condensation.

Cirrostratus: High level (usually around 20,000 ft.) veil-like cloud formations composed of ice crystals and often spreading out over a very large surface area. Halos are very often observed in cirrostratus clouds, indicating a lowering of the cloud ceiling and possible precipitation within 48 hours.

Figure 2-9 Cirrostratus

Altostratus: Medium level clouds (usually hovering around 8,000 ft.) that are flat or striated, dark grey in color. A darkening of the cloud cover indicates possible precipitation within 48 hours.

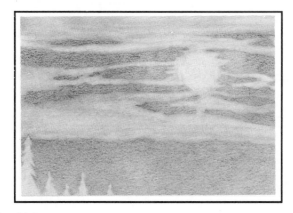

Figure 2-10 Altostratus

Nimbostratus: Low level, dark and thick clouds, often without any real defining shape. Its ragged edges, known as scud, produce steady precipitation.

Figure 2-11   Nimbostratus

Lenticular Clouds: Well known in the Sierra Nevada, lenticular clouds, labeled formally as altocumulus lenticularis, are lens-shaped clouds that form at or around mountain peaks due to a cresting, or wave in the airstream passing over the peak. As air is forced up the mountain ridge, it cools, condenses, and forms into a cloud near the peak. As the air passes over the peak and heads back down the ridge, it warms and moisture evaporates. Since condensation is only occurring at the peak, the cloud forms only at the peak and even though winds are whipping through it, the cloud remains stationary.

Figure 2-12   Lenticular cloud

Anticipating Weather From Cloud Formations

The following are some generally accepted guidelines to help anticipate weather changes by observing cloud movements and sequences.

Cold Front

The rapid formation of altocumulus followed in quick succession by cumulostratus and finally cumulonimbus usually indicates an incoming cold front.

Warm Front

Cloud cover that darkens, beginning with cirrostratus, then altostratus, and finally nimbostratus, over a 24 hour period indicates an approaching warm front and impending rain. Quite commonly, the longer the sequence of cloud formation and darkening, the longer the storm.

Precipitation

If cirrostratus, altostratus and stratocumulus are all visible, the probability of precipitation within 24 hours is high. If the clouds are whipping about in different directions, the precipitation is likely to be very heavy and long in duration. Still having trouble identifying clouds?

Identifying clouds takes practice, practice, and more practice. Get used to systematically working your way through identifications at first and soon, cloud identification will become second nature.

- First, determine if the clouds are heap, layered or precipitating (hint, if you are getting wet, the cloud is precipitating).
- If they are heap clouds only, then estimate the height of the base and top. If the top is no higher than 10,000 feet, then it is fair weather cumulus. If the top is no higher than 25,000 feet then it is swelling cumulus. If the top is above 25,000 feet, then it is cumulus congestus.
- If the clouds are layered only then estimate the height of the base and top. If the cloud is formed between 1,000 to 10,000 feet, it is stratus. If the cloud is formed between 10,000 to 20,000 feet, it is altostratus. If the cloud is formed above 25,000 feet, it is cirrostratus.
- If the clouds are both layered and heaped in shape then estimate the height of the base and top. If the clouds are formed between 1,000 to 10,000 feet, they are stratocumulus. If the clouds are

formed between 10,000 to 20,000 feet, they are altocumulus. If the clouds are formed above 25,000 feet, they are cirrocumulus.

- If the clouds are precipitating (rain or snow) then study the level of precipitation. It does not have to reach the ground to be considered precipitation as is the case with high level clouds, above 25,000 feet, that have an apparent white trail, much like a veil, dropping from the sky, disappearing as it evaporates. This cloud is cirrus. If the showers are steady, fluctuating between light and medium intensity, or just heavy mist, then the cloud is nimbostratus. If the showers are on again, off again with heavy to medium level precipitation, then the cloud is cumulonimbus.

3. GEOGRAPHIC WEATHER VARIATIONS

Although the weather may be warm and sunny on the ridge, it is entirely possible for temperatures to be downright cold in the valley below. How is this possible? Attribute temperature and humidity fluctuations to microclimate influences. The differences in temperatures and conditions in various microclimates are created by the different degrees of solar warming, radiation, and evaporative cooling. Knowing what these are and how they work can make the difference between choosing to spend a chilly night huddled in your sleeping bag within a cold air-sink, or basking comfortably several hundred feet away under moon beams. Of course, if the weather is foul, microclimatology is a mute point— foul weather is a constant with no microclimate sanctuaries.

Adiabatic Fluctuations In Temperature

Air cools approximately 5.5 degrees F for each 1,000 feet of elevation gain. It warms at approximately the same rate during elevation loss. As air meets a mountain and is forced over it, the air pressure reduces and the temperature lowers as the air expands. Once the air pours over the peak and begins to descend, the temperature rises as the air pressure increases and the air mass gets compressed. Adiabatic warming or cooling doesn't only happen on

7000 ft	35.5°
6000ft	41°
5000ft	46.5°
4000ft	52°
3000ft	57.5°
2000ft	63°
1000ft	68.5°
	75°

Figure 3-1   7,000 ft. peak

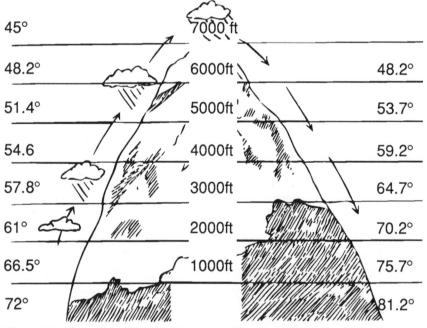

45°	7000 ft	
48.2°	6000ft	48.2°
51.4°	5000ft	53.7°
54.6	4000ft	59.2°
57.8°	3000ft	64.7°
61°	2000ft	70.2°
66.5°	1000ft	75.7°
72°		81.2°

Figure 3-2   Same 7,000 ft. peak illustrating adiabatic cooling and warming.

mountain peaks though. Air rising and lowering through the atmosphere is also subject to adiabatic warming and cooling.

Condensation throws a monkeywrench in the process. Just as adiabatic cooling lowers the temperature of air, any condensation that may occur warms it. The net effect on humid air being pushed up over a mountain peak is an average cooling of the temperature at 3.2 degrees F per 1,000 feet. Once at the peak, the air begins to warm as it descends and no further condensation occurs. Adiabatic warming of this air happens at the normal 5.5 degrees F for each 1,000 feet of elevation loss.

What does all this mumbo jumbo mean? Let's look at a 7,000 foot mountain as an example. As moisture laden air climbs the mountain, it cools and condenses, often creating a localized weather system of rainshowers and mild winds. For the sake of the illustration, we will assume that the condensation and precipitation begins at 2,000 feet. If the temperature at ground level was 72 degrees F, then at 2,000 feet the temperature has been lowered to approximately 61 degrees F (5.5 degrees F for every 1,000 feet. Between here and the peak, the air begins to condense, thus cooling at a slower rate (3.2 degrees F per 1,000 feet) resulting in a peak temperature of 45 degrees F. As the air mass slides over the peak and begins to descend, the condensation evaporates quickly, within 1,000 feet. At 6,000 feet, the temperature has risen 3.2 degrees (the first 1,000 feet were warming at a moist adiabatic rate) to 48.2 degrees. From 6,000 feet down to ground level, however, the air will warm at a dry adiabatic rate since the relative humidity is decreasing, resulting in a final ground temperature of 81.2 degrees F—warmer than when the air mass was forced over the mountain beginning at the same elevation on the other side.

Chinook Winds

As a boy living in Calgary, Alberta, located at the base of the Canadian Rockies, I remember Chinook winds well. This weather phenomena could raise the cold winter temperatures of the morning into a balmy, but very dry afternoon, causing snow to vanish almost magically—confusing indeed to a young boy expecting to go sledding, but ideal sandbox weather nevertheless. Chinook winds occur in the Sierra, the Rockies, and many other mountain locations around North American and the world. In fact, some even carry their own label, such as the famous Santa Ana winds of southern California.

These winds and the accompanying fluctuations in temperature are caused by strong low or high pressure systems that actually force an air mass over a mountain range. As the air mass climbs the mountain, it cools adiabatically as described above, often dropping snow or rain on the windward side of the mountain. Once over the mountain peaks, its moisture depleted, the air mass streams back down the mountain. The warming effect caused by the condensation coupled with adiabatic warming through elevation loss sends hot, dry air roaring through the low lands (temperature climbs of as much as 40 to 45 degrees F within 15 to 30 minute spans have been recorded in Alberta). The winds that blow are fairly constant, usually not over 20 to 25 miles per hour (although winds have been recorded in near the 100 mph mark), and they blow 24 hours a day since they are governed by the driving force of a pressure system, not thermal heating and cooling.

Mountains and Valleys

Hike anywhere in any mountain range and you will note one fairly constant feature, wind blows up the mountain during the day and down the mountain at night. Why? Two forces are at work, both caused by radiation as the earth's surface loses more heat than it absorbs during a clear night. As the ground gives off heat, the heat rises and the earth cools. The air close to the ground becomes colder more quickly and, since it is denser than warm air, begins to flow, rather like water, downhill—towards valley bottoms and desert floors. As the cool air rushes down the mountain (this is the breeze you feel), it displaces the warm air forcing it upward keeping a cycle of air going. As the warm air gets forced out, the cool air pools and collects in the lower reaches, creating places often referred to as cold sinks.

In broad valleys covered with dense meadow grasses, frost pockets may form, even though the temperature just 500 to 1000 feet above is a comfortable 50 degrees. The broader the valley and the more vertical the surrounding mountainous ridges, the more marked the temperature contrasts will be. The narrower the valley, the less dramatic the temperature differences between top and

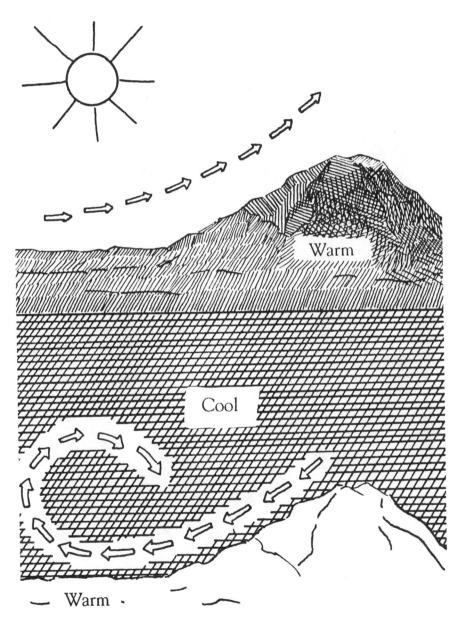

Figure 3-3 Air flow up mountain and down mountain Night; Day.

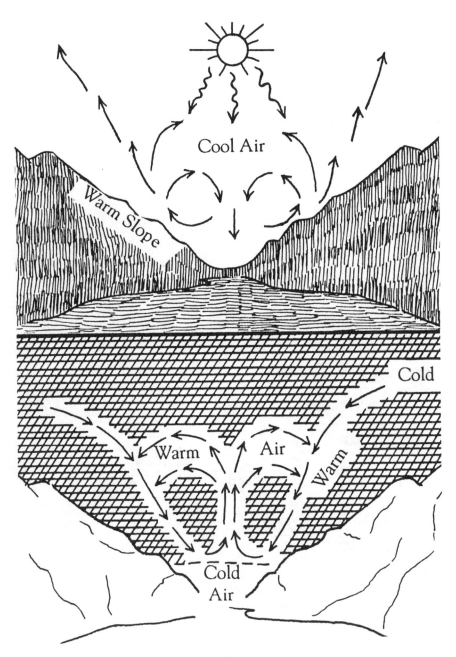

Figure 3-4   Valley air circulation Day; Night.

bottom will be since the walls tend to radiate heat from one to another, effectively trapping potential heat loss from the valley bottom.

As the early morning sun begins to warm the valley, the winds will reverse themselves and begin rushing up the mountain as the warm air begins to rise. Remember, however, that the air can only be warmed if the sun hits the ground below it, radiating heat upwards. This explains why the side of the valley that is getting the sun first will experience upslope winds, while the side of the valley still in the shade continues to experience cool downslope winds.

Cloudy conditions or a presence of strong winds in the area will alter the topographical variations in wind and temperature pattern. Cloudy conditions act as an insulator, reducing radiant heat loss and holding a thermal layer of air closer to the earth's surface. Windy conditions may overpower the gentle nature of upslope or downslope winds and also alter the potential temperature variances.

Winds that flow up and down a valley or canyon floor operate in a similar manner to winds that flow up and down a valley's walls. As temperatures fluctuate between higher elevations (mountains, ridges, or foothills) at one end of the canyon or valley and the lower elevations (desert, plains), winds also fluctuate in intensity and direction. Canoeists and rafters know this well as many an afternoon wind blowing upstream from the heated elevations below has thwarted efforts to paddle into it successfully. Keep in mind, however, that canyons and valleys channel wind too, and it is entirely possible, probable in fact, that winds will roar up or down the canyon or valley floor no matter what time of day it is.

Snowfields and Glaciers

In mountain environments such as the Wind River range in Wyoming, the Cascades of British Columbia, Washington, and Oregon, and the Sierra of California, the presence of year-round snowfields or glaciers create localized wind conditions in the valley's directly below them. These winds are caused by the cool air that flows off the glacier or snowfield, displacing the surrounding warmer air hugging the valley floor. The winds are fairly gentle and warm and dissipate quite quickly, disappearing altogether within a quarter mile of the glacier or snowfield.

Winds originating from glaciers or snowfields will occur as long as there is a sufficient temperature differential between the surface temperature of the snow and the surface temperature of the ground below the lip of the glacier or snowfield. The cold wind that emanates from ice caves underneath a glacier also has a significant effect on the growing season of plants. In the Cascade range near the Canadian border, there are several ice caves that I have camped below where the growing season of plants and flowers consistently begins several weeks after those plants only several hundred feet away. This knowledge works well if you are trying to get a photograph of various flowering alpine plants as I was, and miss the first bloom due to miserable photography weather, as I did.

Alpine Environments

Wind is a constant in alpine environments, chapping lips, dehydrating the body, cooling the skin and generally governing how and where plants will grow. Local upslope winds, caused by the sun heating the earth in the lower elevations, begin around mid-morning. As the sun sets, the reverse effect is true as downslope winds begin, caused by the rapid cooling of alpine soils. Backpackers and mountaineers know not to camp in or near the bottom of sub-alpine valleys because of the intense cold that can be expected there—far colder than the tundra surface above.

Wind patterns in alpine environments can be altered by variations in the mountain terrain—rock outcrops, steep rock faces or slopes, and narrow passes or valleys. These variations in terrain help to create localized wind gusts, swirling eddies, and/or a phenomena known as Venturi effect. As wind gets forced through a narrow constriction, the localized air pressure drops and the wind accelerates. There is a pass in the Wind River Mountains where the prevailing wind on either side is relatively gentle. In the pass, however, the wind is felt almost gale force, blowing a tiny waterfall back up a rock face, evaporating it before it ever has a chance to hit the ground.

Deserts

Thermally created low pressure systems often develop in desert areas. As super heated air rises skyward, it creates a void, a

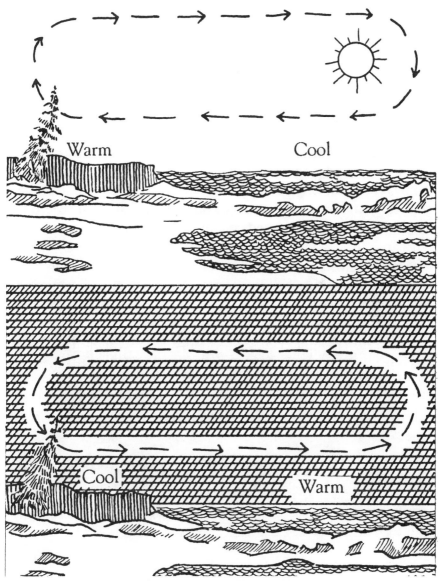

Figure 3-5 Land and sea air circulation differences between Day; Night.

low pressure area that must be filled. The resulting effect is strong horizontal winds rushing in as the pressure attempts to equalize, creating dust devils and sandstorms.

Lakes and Oceans

Anyone who has spent anytime at all hiking or boating near the shore of a large body of water, such as the ocean or any one of the Great Lakes, is probably familiar with the daytime onshore winds and the nighttime offshore breezes. As with snowfields and glaciers, temperature differences between the land and sea must be sufficient for winds to occur. Unlike snowfields and glaciers, clouds will reduce the temperature contrasts. Winds typically are most common when the days are sunny and warm and the nights are clear.

Generally, on shore breezes will begin around mid-morning and continue until early to mid-afternoon. The wind will typically be steady and can result in a dramatic drop of on-shore temperature, sometimes as much as 10 to 15 degrees F. Evenings and early mornings are usually calm. The off shore breeze often begins shortly after dusk and continues until early morning.

City Environments

Cities create their own microclimates that are unique to them. Because of the high buildings, the congestion of people, and the

Figure 3-6   Thermal Mountain

concentrated production of pollutants, city air and weather is often dramatically different than the urban air only a few miles away.

As the buildings absorb the sunlight, it traps the heat close to the ground. In addition, the particles of pollution, hanging over the city, serve to create a kind of seal, holding the heat in and not letting it escape into the atmosphere. This is why city temperatures are often 5 to 10 degrees F warmer than the surrounding countryside. This phenomena has been named by some, the Thermal Mountain.

Cities also generate a great deal of moisture laden air which often leads to rainfall and fog within city boundaries, but not in the surrounding countryside.

General Guidelines

Microclimates are as varied as the topography they occur in, but some generalities can be drawn.

Wooded terrain is generally warmer in the winter and cooler in the summer than open, treeless areas. Winds are lower and the humidity is higher in the woods when compared to open areas.

Valleys have lower daily temperatures, wider temperature fluctuations, more frost and dew accumulation and more fog frequency than higher up in the hills or mountains. Wind velocity is also less in valleys when compared to higher elevations.

4. FORECASTING CHANGES IN WEATHER

The Role of an Altimeter to Observe
Barometric Pressure Changes

An altimeter is, in essence, an aneroid barometer, capable of reading relative changes in barometric pressure and, with the addition of scale inscribed on the face, changes in elevation. Because of its compact nature, an altimeter is not ideally suited for reading the minute changes (recorded in 1/100 of an inch) used by the National Weather Service, but the standard 1/10 of an inch increments are adequate for amateur use.

The innards of an altimeter are quite basic, consisting of a vacuum chamber that contracts with high pressure (those pressures found closer to sea level) and expands with low pressure (pressures get lower the higher one climbs). Through the intricacies of gears and gizmos, the information is transferred and displayed by a needle on the face of the altimeter.

In general, when the weather service refers to a falling barometer, a lowering of the atmospheric pressure, the altimeter shows a rise in elevation. Conversely, a rising barometer, a rise in the atmospheric pressure, is read as a falling elevation on the altimeter.

How does this relate to changes in weather? Without getting specific, a rise in barometric pressure, demonstrated by a drop in your

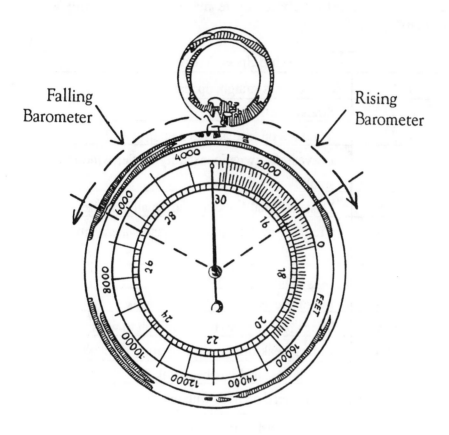

Figure 4-1 Altimeter

elevation reading while camped in one location, indicates improving or continued good weather. A drop in barometric pressure or a rise in elevation readings is a precursor to deteriorating weather.

Be warned that any rapid change in barometric pressure, up or down, promises a change in the weather, and not always for the best. Slow changes usually indicate a stable weather pattern (dry or wet) lasting for a while.

Estimating wind speed

Devised by Sir Francis Beaufort in 1805, the wind speed chart, now known as the Beaufort Scale of Wind Force, is an excellent, and extremely accurate method of estimating wind velocity.

	Wind Speed MPH
0-1	Smoke rises straight up, calm
1-3	Smoke drifts
4-7	Wind felt on face, leaves rustle
8-12	Leaves and twigs constantly rustle, wind extends small flag
13-18	Dust and small paper raised, small branches moved
19-24	Crested wavelets form on inland waters, small trees sway
25-31	Large branches move in trees
32-38	Large trees sway, must lean to walk
39-46	Twigs broken from trees, difficult to walk
47-54	Limbs break from trees, extremely difficult to walk
55-63	Tree limbs and branches break
64 on up	Widespread damage with trees uprooted

Wind Chill Index

	Temp	40	30	20	10	0	-10	-20	-30
Wind MPH	5	37	27	16	6	-5	-15	-26	-36
	10	28	16	2	-9	-22	-31	-45	-58
	15	22	11	-6	-18	-33	-45	-60	-70
	20	18	3	-9	-24	-40	-52	-68	-81
	25	16	0	-15	-29	-45	-58	-75	-89
	30	13	-2	-18	-33	-49	-63	-78	-94
	35	11	4	-20	-35	-52	-67	-83	-98
	40	10	-4	-22	-36	-54	-69	-87	-101

The wind chill equivalency index chart, shown on the preceding page, demonstrates the actual cooling effect on bare skin when exposed to air. Although understanding wind chill has very little to do with being able to predict weather patterns, it is important to realize the impact wind and temperature have on the body in order that the outdoorsman or woman can make safe decisions regarding dress and travel.

Putting It All Together

Combine the altimeter with wind direction, temperature and cloud conditions to predict weather. Some generalized examples are as follows, although keep in mind that if predicting the weather were as simple as the following examples appear to make it, weather forecasters would never be wrong.

Warm Front

Incoming Storm

The barometric pressure will fall steadily (the altimeter elevation rises). Wind is out of the SE or NE and increasing in speed. Cirrus will give way to altostratus, then nimbostratus. As the clouds begin to thicken, precipitation begins to fall, increasing in intensity as the front moves through. Temperature will increase gradually.

Outgoing Storm

The barometric pressure will level off. Wind direction changes to come more from the S or NW. Nimbostratus gives way to stratocumulus. Precipitation lightens to showers and drizzle.

Cold Front

Incoming Storm

The barometric pressure will fall (altimeter elevation rises) slowly at first, more rapidly as the storm approaches. Wind will be from the S or NE. Cumulus gives way to cumulonimbus. Brief but intense showers or hail. Little change in temperature.

Outgoing Storm

The barometric pressure will rise sharply (altimeter elevation falls). Wind direction will shift or the N or NW and become gusty. Cumulonimbus will begin breaking up with partial clearing. Showery with intermittent thunderstorms followed by rapid clearing. Temperature will drop rapidly.

Occluded Front

Incoming Storm

Barometric pressure falls steadily. Winds typically from the E or NE, sometimes from the SE with velocity increasing. Cirrus giving way to cirrostratus, then altostratus, then nimbostratus. Steady precipitation increasing as storm moves through. Temperature rises slowly.

Outgoing Storm

Barometric pressure rises steadily. Wind from the SW or N and decreasing. Stratocumulus to altocumulus with slow clearing. Precipitation tapering off. Temperature falls slowly.

Learn to use the wind direction and your reading of the altimeter's fluctuations of elevation when stationary to aid in anticipating possible upcoming weather changes.

Using Wind Direction to Predict Weather

The following information has been adapted from data regarding average conditions within the United States and obtained from the National Weather Service.

Winds from the NW

When winds are from the NW and the weather has been clear you may anticipate 24 hours of continued clear weather if your altimeter is steady or shows a slow drop in elevation (a rise in barometric pressure). If it has been raining and you read a drop in elevation, weather should clear in several hours. In all cases, expect lower temperatures.

If the altimeter shows a rise in elevation (a drop in barometric pressure) expect it to remain clear for 24 hours if it has been clear or expect the weather to change if it has been stormy.

Winds from the SW

When the winds are from the SW and the weather has been clear, you may anticipate continued fair weather for 12 to 24 hours if your altimeter shows a drop in elevation (a rise in barometric pressure). If it has been stormy and you read a drop in elevation, the storm should pass in 6 hours.

If the altimeter shows a rise in elevation (a drop in barometric pressure) you may predict rain within the next 12 hours if the weather has been clear, or expect increasing rain with possible clearing after 12 hours if the weather has been stormy.

Winds from the SE

If the weather has been clear and the winds are from the SE, you may plan on continued fair weather if the altimeter shows a drop in elevation (a rise in barometric pressure). If the weather has been stormy, anticipate clearing weather to follow soon if the altimeter shows a drop in elevation.

A rise in elevation shown on the altimeter during clear weather would indicate impending rain within 12 hours and possible high winds. If the weather is already stormy, a rise in elevation indicates an increase in severity of the storm followed by clearing within 24 hours.

Winds from the NE

When the winds are from the NE and the weather is clear, expect continued fair, but cool weather if the altimeter shows a drop in elevation (a rise in barometric pressure). If the weather has been stormy, expect it to clear the the temperature to drop.

If the altimeter rises in elevation while the weather is clear, expect stormy weather within 12 to 24 hours. If the weather is stormy, expect very heavy rain coupled with severe gale force winds and much colder temperatures.

5. NATURE'S SIGNS

semi-reliable forecasting from legends and lore

Through the centuries, man has made devoted efforts to study and anticipate weather changes. Long before the advent of sophisticated weather paraphernalia, man studied weather by reading all the available signs. He listened to the animals, the plants, the wind, the earth. He used his eyes, his ears, his nose and taste. Not surprisingly, records show that these "ancient ways" are almost as reliable as modern methods. Arm yourself with weather tools if you will, but don't ever close your eyes and ears to the signs of nature all around.

Morning or Evening Sky

"Red sky at night, sailor's delight. Red sky in morning, sailors take warning." This means that if the clouds take on a reddish hue in the morning, one can expect rain by the end of the day. If the evening sky is red, the weather will probably remain clear the following day.

Geese

Geese have a reputation for not flying before a storm. Biologists theorize that this is because they have a harder time getting airborne in low pressure (thinner air) conditions. Or, maybe its just that they are a lot smarter than we give them credit for.

Figure 5-1

Coffee

Grizzled outdoorsmen and women swear by the bubbles in the coffee method. Perhaps this explains why so many of them spend hours staring into a steaming mug of Java? Remarkably, it does seem to work, attributed to the way pressure affects the meniscus, or surface tension of the coffee. In high pressure, the surface is rounded, like a globe. In low pressure, the surface is concave, so naturally bubbles head to the highest point on the coffee's surface, the edges of the cup. For coffee forecasting to work, the brewed coffee must be strong. Instant won't do it as there doesn't seem to be enough oils to create satisfactory surface tension. Pour the coffee into a mug (vertical sides work best; venerable Sierra cups don't work as well). Give the coffee a good stir or two and watch the bubbles form. If they scatter this way and that and then form near the center, fair weather. If they cling to the sides of the cup, a low pressure system is setting in and rain is possible.

Mosquitoes and Black Flies
Anyone who has camped beside a body of fresh water prior to a storm's arrival knows that mosquitoes and black flies will inevitably swarm and feast heavily up to 12 hours before the storm hits. You can bet the storm is almost upon you when the feasting frenzy subsides as the little buggers seemingly fly away to hide approximately one hour before the weather turns.

Fog
If the dawn is gray and there is fog in the valleys, the weather that day will be clear.

Frogs
Frogs of all shape and color have an uncanny tendency to increase their serenading several hours before a storm arrives. The reason they do this is not so much their reliability as a weather forecaster, but because the increased humidity in the air from an incoming storm allows them to stay comfortably out of the water for longer periods — their skin must be kept moist at all times.

Bees
A friend of mine, who is an amateur bee keeper, once told me that bees can sense inclement weather and will stick closer to the hive when the weather is about to take a turn for the worse.

Springs, Ponds and Caves
Natural springs seem to flow out of the ground at a higher rate when a storm is approaching. This has been attributed as a natural response to a lowering of the barometric pressure.

The same lowering of barometric pressure will cause ponds, those with a lot of vegetative decay at their bottom, to become momentarily "polluted" as the decaying scum rides to the surface on marsh gases.

Stand near a cave's entrance during a storm's approach and you will become very aware of an outrushing of air. This is also due to a lowering of barometric pressure, as the air pours out of the cave to equalize the outside pressure. Just because a cave is breathing, however, doesn't mean a storm is on the way. Caves often breath naturally in response to outside temperatures.

Hair
Although not everyone is affected in quite the same manner, and women tend to notice the phenomena more frequently than

men (possibly because they are more aware of their hair?), hair does react to humidity changes. Because hair is a reliable indicator of good or bad weather, some instruments that measure humidity, called hygrometers, use hair as the primary working element. Like rope, hair tends to contract when it is damp and relax when it is dry. Straighter hair means dry weather. Wavier or curlier hair means wetter weather.

Halo Around the Sun and/or Moon
In the summer, the sight of a hazy halo or corona around the sun or moon is a good indication that a change in the present weather pattern is in the forecast, most often for rain.

Frost and Dew
The presence of heavy frost or dew early in the morning or late in the evening is a fairly reliable indicator that up to 12 hours of continued good weather may be expected.

Canvas, Hemp, Wood Axe Handle Changes
Like human hair, hemp and canvas contracts when it gets humid and stretches as it dries out. Woodsmen in the olden days noticed humidity changes as their wood ax handles swelled during higher humidity (incoming rain) and loosened during drier weather.

Sea gulls
Sailors know that if sea gulls are clustering the beaches and not flying about over the ocean, it is a good indication that a fairly strong storm is about to blow in and boats had best stay in the harbor.

Song Birds
There is a theory that some birds tend to sing loudest right before a storm's arrival. Of course, this assumes that you have been carefully monitoring the volume of the birds in your area all day long. There is an equally popular theory that song birds become silent right before a storm. I guess the real dilemma is which songbirds are going to sing what song and when. My suggestion is to take your pick, but keep a weather eye skyward just in case.

Colors of the sky
Tints and hues of green, yellow, dark red or a grayish-blue indicate precipitation and quite often accompanying winds.

Cattle

Farmers have for years looked to their cattle to give them reliable indications of changing weather. It seems that cattle will herd together in lower elevations and off exposed hills when the weather is about to take a shift for the worse. To some extent, the same is true of deer and other grazing animals.

Wind

"Wind from the south brings rain in its mouth." If you look back to Chapter one and the discussion regarding winds in the Northern Hemisphere, you will note that low pressure systems create cyclonic winds that rotate in a counterclockwise direction. Since low pressure systems are frequently associated with rainstorms, the rhyme proves quite accurate. Counterclockwise wind rotations create wind that blows from the south, wind that brings in the rain.

High pressure systems are often associated with clear or clearing weather, with clockwise rotating winds. Keep aware of wind directions and you will keep your finger on the weather's pulse—is it beating fair or foul?

Another theory that holds a certain amount of credibility states that if the wind comes first from the north then the west and finally the south, rain is imminent. Winds beginning in the south, shifting to west and then back to north indicates the storm or cloud cover is clearing.

Smell

In the Midwest and the Great Plains, the smell of rain, has a very definite and understood meaning. As a storm approaches, the pressure drops and the humidity level rises, the ground begins to give off a very rich and sweet odor that almost smells like freshly mown hay.

Sound

Sound does, apparently, travel further when a storm approaches. Higher humidity helps to carry sound waves, as does the wind from an approaching low.

Plants

Dandelions are often credited with being cold front predictors as they will close up their flowers as the temperature drops below 50 degrees F.

Observe a field of clover and you will note that the clover rolls its leaves up at the approach of strong winds. Since strong winds often precede a storm, clover is known as a reliable indicator of incoming weather. Still, one could argue that it is the wind and not the clover that is the harbinger of weather tidings.

Campfire Smoke

By observing the smoke from your campfire you can observe what pressure system is in the area, low or high. If the smoke from the fire hangs low to the ground and dissipates into the branches, that

Figure 5-2a

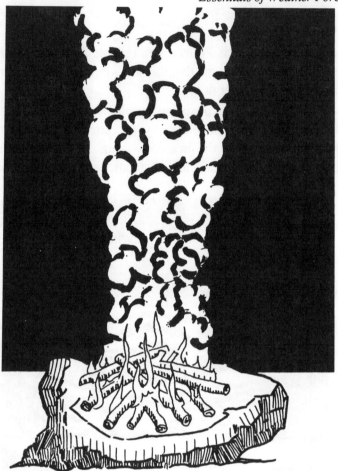

Figure 5-2b

means a low pressure system is here and rain is possible. If the smoke rises in a straight, vertical column, high pressure rules and fair weather may be anticipated.

The Calm Before a Storm

If the wind has been blowing and clouds moving in rapidly for the past few hours and then, suddenly, the wind dies, head for cover because you are in for one heck of a gully washer.

6. BACKYARD METEOROLOGY

Setting up a backyard weather station is not only fun, but it can be useful as well. Combine a little experience and the right equipment, and anyone can learn to safely predict weather changes at least 12 to 24 hours in advance, and that can make the difference between knowing whether to pack a raincoat or sunscreen on your next jaunt around town.

With the increasing use of computerization, home weather stations have become more compact and certainly more high tech than ever before. One such station, the Weather Wizard II, manufactured by Davis Instruments of Hayward, California, allows the amateur observer to document all pertinent weather information from the comforts of his/her home, without ever stepping a foot outdoors. A miniature computer registers information regarding wind speed and direction, temperature, barometric pressure changes, and rainfall obtained from outside instrumentation mounted on the roof or other open area. Talk about your cozy weather forecasting. On the other hand, one could argue that all the computerization could lead to a decidedly detached approach to forecasting—nothing like being in the thick of things for true "hands on" and experiential weather observation.

For a more traditional approach, you will want to create a weather station that is out in the open, with all the instruments close enough together to facilitate easy reading and recording of information. To get the most out of a weather station, you will want to assemble the following instruments: barometer, anemometer, wind vane, and thermometer. A rain gauge and hygrometer are accessories that, while highly informative, are not absolutely necessary for basic weather forecasting. They are most useful, however, if you wish to assemble a complete weather station and have the space.

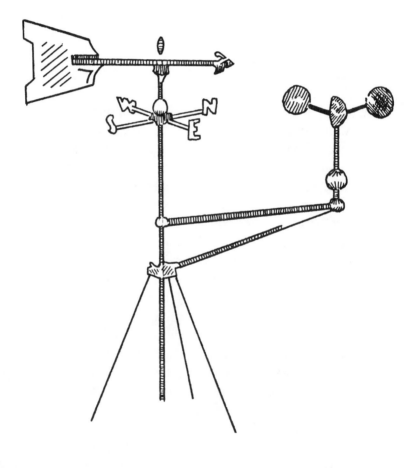

Figure 6-1   Weather vane and an anemometer

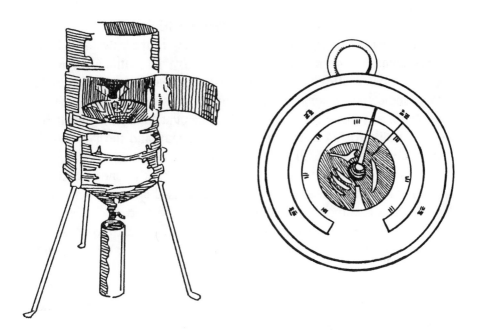

Figure 6-2   A rain gauge and barometer

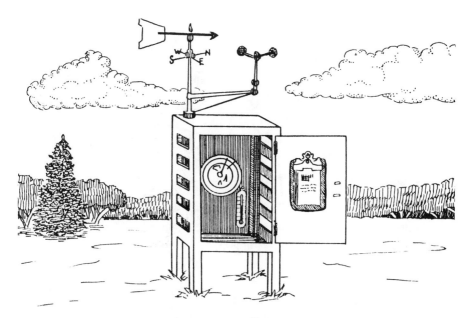

Figure 6-3   A weather station. A box on stilts away from buildings, the sides well ventilated (slotted) and weather gear (barometer and notebook) inside and wind vane and anemometer mounted on top.

Understanding a Weather Map

The weather maps that you see every day in the newspapers are usually created from information provided by the National Weather Service maps. While the newspaper weather maps are not the most reliable indicators of long-range weather forecasts, with practice, it is possible for anyone to become somewhat adept at anticipating what the weather holds for 24 hour periods. USA Today and the New York Times offer the best overall reproductions of a weather map.

In order to read a weather map, it is important to understand the symbols. The thin black wavy lines that cross the map, resembling contour lines common on topographic maps, are isobars. Isobars are lines of equal barometric pressure. These lines illustrate pressure patterns that control air flow. Air moves fastest where the lines are closest.

Four other lines on a map have geometric symbols on them—one all triangles, one arches, one an alternating combination of triangle and arches facing the same direction, and the other an alternating combination of both triangles and arches facing opposite directions. All

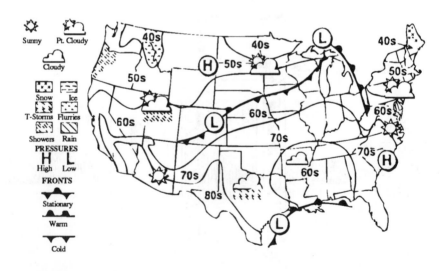

Figure 6-4   Weather Map

triangles indicates a cold front. All arches indicates a warm front. The alternating combination of triangle and arches facing the same direction indicates an occluded front. The alternating combination in opposite directions shows a stationary front, a front that is not moving. The direction either a cold or warm front is moving is indicated by the direction the triangles or arches are pointing.

Wind directions and speeds are designated by tiny symbols that look remarkably like musical notes—a little circle with a tiny flag attached. The arm or flag of the symbol points in the direction the wind is coming from.

Shaded areas of the map indicate areas where precipitation is falling. Check the map's key to determine how rain or snow is differentiated.

Developing a Routine

Every day, at the same time each day, establish a weather observation routine that never varies. You don't need to spend more than 10 to 15 minutes each time, but consistency is the key to acquiring accuracy. Be sure to document your observations and weather comments in a log book. That way, you can go back and establish a weather history, learn the patterns that lead to various weather systems.

1. Look to the sky. Observe and identify as many cloud formations as you can. Note the direction they are moving and whether or not they appear to be part of a system. Are they forming, or breaking up? What is their sequence of formation?

2. Using your weather vane and anemometer (or rely on the Beaufort scale for estimating) determine the wind speed and direction. Has it changed from yesterday? How? Is the wind increasing or decreasing? Does the wind direction, coupled with the formation sequence of clouds indicate a front moving in?

3. Check the temperature and the barometric pressure. Is the temperature going up or down? Why? Is the barometric pressure going up or down or is it holding steady? Why? How does this information mix with what the cloud formations tell you?

4. Check the rain gage for precipitation levels.

5. Enter all the information in your log. Your log can be set up anyway you want, but it should include separate columns for the following: Date, time, wind direction and velocity, temperature, barometric pressure, precipitation, forecast, changes. Changes are for you to fill in after your forecast if what you predicted doesn't pan out. Don't forget to note why what you predicted didn't happen and what happened to the weather to change the conditions. It is also a good idea to have a special column where you can note conditions such as frost, dew, or fog.

APPENDIX I
Glossary of Terms

Acid Rain: Rain that contains acid pollutants produced by factories and automobiles. Acid rain leaves lakes infertile, unable to support fish or plant life. It also kills vegetation and trees.

Adiabatic Cooling: Cooling of the air due to a rise in elevation.

Adiabatic Warming: Warming of the air due to a lowering in elevation.

Anemometer: Instrument for measuring the velocity or miles per hour that the wind is traveling. It is important to note that most weather bureaus report the wind speed in nautical miles per hour or knots as this is a universally understood unit of measurement.

Atmospheric Pressure: The weight of a column of air on the earth's surface measured by a barometer in inches or centimeters.

Barometer: The instrument used to measure atmospheric pressure.

Chinook Winds: Relatively strong winds, usually warm and dry, that occur in valleys and canyons on the leeward side of mountain ranges.

Coalescence: The creation and growth of a droplet of rain large enough to fall to earth through the atmosphere.

Coriolis Effect: The affect on air movements caused by the earth's rotation. Air movement curves to the right in the Northern Hemisphere and to the left in the Southern.

Corona: A corona may be viewed when the light from the sun or moon passes through a thin veil of high moisture-laden clouds. Most often, the corona looks like a yellowish/white disk surrounding the moon or sun, sometimes with an inner blue border and a dirtier brown outer edge.

Dew Point: Dew point is the temperature at which the relative humidity equals 100%. This causes moisture in the atmosphere to condense into rain.

Downdraft: Flow of air downward due to a lowering in air temperature most often caused by falling precipitation.

Drizzle: A very light rain fall. Droplets are uniform in size, quite small and typically fall from stratus clouds.

Front: The boundary between two air masses, named for the dominant air mass moving in—cold, warm or occluded front.

Graupel: A mass of frozen droplets of precipitation that is lumpy and soft. Frequently falls in conjunction with a severe thunderstorm. Softer than hail, more bulky than snowflakes.

Green Flash: A seldom viewed and somewhat mystical occurrence at sunset that is seen immediately following the disappearance of the red rim of the sun beneath a clearly defined horizon. The green flash is created by sunlight refracted through various densities of atmospheric layers and lasts only a few seconds.

Hail: Forming most typically in large, billowing cumulonimbus clouds. Moisture particles freeze in areas of intense updrafts, often falling earthward only to be tossed up once again by a strong updraft. Large hail is formed as hail particles begin to thaw, then refreeze after gathering more moisture. Hail stones as large as baseballs have been recorded.

Halo: A halo may sometimes be viewed surrounding the sun or moon and is caused by the presence of ice crystals in cirrostratus clouds that fall between the light source and the observer.

High Pressure: An air mass with greater than normal air pressure, often referred to as an anticyclone.

Hygrometer: Instrument used to measure the amount of moisture in the air otherwise known as humidity.

Isobars: Lines on a weather map that indicate lines of equal barometric pressure.

Jet Stream: The name given a band of rapidly moving air caused by cold polar air moving south and warm tropical air moving north. The jet stream heavily influences storms and is frequently referred to by weather forecasters. The jet stream is most powerful in the winter since global temperature contrasts are at their greatest.

Low Pressure: An air mass with lower than normal air pressure, often referred to as a cyclone.

Mist: Extremely small raindrops falling at a steady rate from stratus clouds.

Orographic lift: When physical barriers such as mountain ranges force air currents to rise, this is known as orographic lift.

Ozone: At the top of the atmosphere, a shell, called the ozone layer shields the earth from ultraviolet rays. Ozone depletion, or a weakening of that layer, is attributed to chlorofluorocarbons which are used in refrigerators, air conditioners, industrial solvents, spray can propellants, and the manufacture of insulation and polystyrene foam.

Rain: Larger than drizzle yet smaller than showers, rain is made up of consistently sized droplets of moisture that fall steadily from nimbostratus clouds.

Rainbow: When the sun's rays pass through a water drop, the light that passed through the extreme upper and lower edges of the drop bend by refraction creating the colors the eye sees as a rainbow.

Raindrop: Contrary to popular myth, raindrops are not shaped like a tear drop. Instead, they are round in shape when small and somewhat flatter in shape when large. The shape is determined by friction with air.

Relative Humidity: The amount of measurable moisture in the air.

Rime: Usually only viewed on exposed rock faces and other objects at or near mountain summits. Originating from a supercooled cloud passing rapidly over a mountain summit, droplets freeze upon impact creating wonderfully beautiful ice formations that take on a feather-like quality. Ice formations up to several feet long protruding from tree branches and rock faces are not unusual.

Saturation Point: Also known as 100% humidity, saturation point is the point at which the air cannot hold anymore moisture.

Showers: Made up of the largest of the droplets of moisture falling from clouds, shower droplets are large in size and fall in dense sheets from heap clouds such as cumulonimbus.

Sleet: Precipitation that is partially frozen and partially liquid. Formed predominantly when drizzle falls through cold air and begins to freeze from the outside first. The stinging sensation of wet ice against skin is not a pleasant one.

Snow: Frozen precipitation that may take many crystalline forms depending greatly on the temperature and moisture content of the clouds in which the snow formed.

Sun dog: This is the term for unusual bright spots, or second suns, that can appear on both sides of the sun through refraction involving the viewer, the sun and ice crystals in the air.

Thermal Mountain: The invisible dome of hot air that builds over a city environment, prevented from escaping by the particles of dirt and pollution that hang over the city.

Thermometer: Instrument used to measure fluctuations in air temperature.

Updraft: An upward flow of air from a warm surface, usually caused by the earth's surface being warmed by the sun.

Venturi effect: Wind is forced through a constriction (a narrow mountain pass, narrow streets between skyscrapers) causing a drop in air pressure and an associated acceleration of wind.

APPENDIX II
Recommended Reading

American Weather Observer, a monthly publication of the Association of American Weather Observers; P.O. Box 455, Belvidere, IL 61008

Making and Using Your Own Weather Station, by Beulah and Harold Tannenbaum; published by Franklin Watts, 1989. A superb book providing instructions on how to build your own weather instruments and how to use them to observe and forecast weather.

National Weather Service Publications
Various useful publications may be ordered from the National Weather Service by writing the Superintendent of Documents, U.S. Government Printing Office, Washington, D.C. 20402. Ask initially for a listing of publications and the associated price list. Some of the more useful pamphlets that were in print at the time of publication are: *Climates of the United States; Cloud Code Chart; Instruments used in Weather Forecasting; Weather Forecasting.* a particularly outstanding book on weather, also available from the U.S. Govt. Printing Office at the above address is, *Aviation Weather for Pilots and Flight Operations Personnel,* coauthored by the FAA and the Department of Commerce.

Peterson First Guides, Clouds and Weather, by John A. Day and Vincent Schaefer; published by Houghton Mifflin Company, 1991. A pocket guide to understanding weather.

Pocket Weather Forecaster, by Weather Trends Inc.; published by Barrons, 1974. Pocket guide allowing the user to predict weather patterns by reading cloud formations and wind direction and then using a sliding chart to compute the information and arrive at a forecast.

The Weather Companion, by Gary Lockhart; published by John Wiley & Sons, Inc., 1988. A marvelous and detailed look at meteorological history, science, natural legends and folklore.

Weathering the Wilderness, by William F. Reifsnyder; published by Sierra Club Books, 1980. An excellent guide to specific weather patterns and detailed weather information for regional wilderness climates.

APPENDIX III
SUPPLIERS OF WEATHER INSTRUMENTS

Davis Instruments, 3465 Diablo Ave., Hayward, CA 94545

Radio Shack, One Tandy Center, Fort Worth, TX 76102

Taylor Scientific Instrument, 95 Glenn Bridge Road, Arden, NC 28704

Wind and Weather, P.O. Box 1012-W, Mendocino, CA 95460

APPENDIX IV
WEATHER RULES OF THUMB

Clip out the following weather guidelines and take them with you. Combined with all the observation techniques and skills you are acquiring, they will give you a solid foundation from which to make accurate predictions concerning weather trends over any 24 hour period. Add your own notes as you wish.

It is going to be fair when:
- Wind is blowing from the west or northwest
- The barometric pressure remains steady or rises slowly
- Fair weather cumulus clouds dot the sky
- Early morning fog evaporates (gets burned off) by noon

It is going to rain or snow when:
- The barometric pressure falls
- Cumulus clouds begin to develop vertical columns
- There is a halo around the moon
- Cirrus clouds begin to thicken and the cloud ceiling lowers
- The sky darkens
- A south wind increases in velocity
- A wind shifts in a counterclockwise direction, indicating a low.

61

A north wind shifting to west and then to south is a prime example

Expect the weather to clear when:
- The barometric pressure begins to rise quickly
- South winds shift to the west
- Cloud ceiling begins to lift

Anticipate cold temperatures when:
- The night is clear, no cloud cover, and there is virtually no wind
- The barometric pressure rises in the winter or ahead of an oncoming system of clouds, a cold front

INDEX

Adiabatic Fluctuations (cooling or warming): 14, 23, 24, 25, 26
Alpine Environments: 30
Altimeter: 4, 34, 35, 38, 39
Altocumulus: 16
Altostratus: 4, 5, 6, 19
Anemometer: 48, 49
Aneroid Barometer: 34, 49
Atmospheric Pressure 9, 34
Barometric Pressure: 2, 3, 4, 34, 35, 38
Beaufort Scale of Wind Force: 36
Bees: 42
Campfire: 45
Canvas: 43
Cattle: 44
Chinook Winds: 25
Cirrocumulus: 15, 16
Cirrostratus: 4, 19, 21
Cirrus Clouds: 4, 15
City Environments: 32
Clover: 44
Coalescence: 10

Coffee: 41
Cold Front: 3, 4, 5, 14, 21
Condensation: 10, 14, 20, 25, 26
Continental Polar: 1, 2
Cumulonimbus: 5, 6, 17, 21
Cumulostratus: 5, 21
Cumulus: 16
Cumulus Congestus: 18
Dandelions: 44
Deserts: 30
Dew: 10, 33, 43
Dew Point: 4
Downdrafts: 7
Evaporation: 13, 14
Fair-weather Cumulus: 17
Fog: 33,42
Frogs: 42
Fronts (see warm front, cold front, occluded front)
Frost: 10, 33, 43
Geese: 40
Glaciers: 30
Hail: 17
Hair: 42
Halo: 19, 43
High Pressure: 2, 34, 41, 46
Hurricanes: 9
Incoming Storms: 37
Lakes: 32
Lenticular Clouds: 20
Lightning: 6, 7, 8, 17
Low Pressure: 9, 30, 34, 41, 46
Mares Tails: 15
Maritime Polar: 1
Maritime Tropical: 1
Microclimates: 23
Mosquitoes: 42
Mountain: 20, 23, 26, 29
National Weather Service: 9, 34, 38

Nimbostratus: 4, 5, 20, 21
Occluded Front: 3, 5, 38
Oceans: 9
Outgoing Storm: 37, 38
Precipitation: 3, 7, 9, 14, 21, 25
Radiation: 10, 26
Rain Gauge: 48, 49
Santa Anna: 26
Scud Clouds: 15
Sea gulls: 43
Seasons: 10, 11, 12
Smell: 44
Snow: 9
Snowfields: 29, 30
Song Birds: 43
Sound: 44
Springs: 42
Stratus: 19
Swelling Cumulus: 18
Thermal Layer: 29
Thermal Mountain: 33
Thermometer: 48
Thunder: 7, 17
Thundershowers: 5, 6, 7, 16
Tornados: 9
Updrafts: 6, 9
Valley: 26
Venturi Effect: 30
Warm Front: 3, 14, 21, 37
Water Cycle: 13
Weather Map: 50
Weather Station: 47, 48
Winds: 38, 39, 44
Wind Chill Index: 36
Wind Vane 48, 49